PHILOSOPHICAL PHYSICS

THOUGHTS COMPARING SCIENCE

T S KEERTHANA

Copyright © T S Keerthana
All Rights Reserved.

This book has been self-published with all reasonable efforts taken to make the material error-free by the author. No part of this book shall be used, reproduced in any manner whatsoever without written permission from the author, except in the case of brief quotations embodied in critical articles and reviews.

The Author of this book is solely responsible and liable for its content including but not limited to the views, representations, descriptions, statements, information, opinions and references ["Content"]. The Content of this book shall not constitute or be construed or deemed to reflect the opinion or expression of the Publisher or Editor. Neither the Publisher nor Editor endorse or approve the Content of this book or guarantee the reliability, accuracy or completeness of the Content published herein and do not make any representations or warranties of any kind, express or implied, including but not limited to the implied warranties of merchantability, fitness for a particular purpose. The Publisher and Editor shall not be liable whatsoever for any errors, omissions, whether such errors or omissions result from negligence, accident, or any other cause or claims for loss or damages of any kind, including without limitation, indirect or consequential loss or damage arising out of use, inability to use, or about the reliability, accuracy or sufficiency of the information contained in this book.

Made with ♥ on the Notion Press Platform
www.notionpress.com

AUTHOR BIOGRAPHY:

I am Ms. T. S. Keerthana, daughter of Mr. R.Senthil kumar and Mrs.N.Jeya kumari. The only persons whom were being my everything as loyal in this world. And one of my inspirations, my sister S.J. Dhiviya who truly believes me without giving up. Here I am the Physics student who always don't judge anybody rather I wish to learn good from anyone by ignoring flaws. I am the ardant fan of Mr.Feynman, the father of quantum mechanics. Being a Physics student, I love his lectures.

Contents

Foreword   *ix*

Preface   *xi*

Acknowledgements   *xiii*

Prologue   *xvii*

1. Adage   1
2. Adage   4
3. Adage   6
4. Adage   8
5. Adage   10
6. Adage   12
7. Adage   14
8. Adage   16
9. Adage   18
10. Adage   20
11. Adage   22
12. Adage   24
13. Adage   26
14. Adage   28
15. Adage   30
16. Adage   32
17. Adage   34
18. Adage   36
19. Adage   38
20. Adage   40

Contents

21. Adage	42
22. Adage	44
23. Adage	46
24. Adage	48
25. Adage	50
26. Adage	52
27. Adage	54
28. Adage	56
29. Adage	58
30. Adage	60
31. Adage	62
32. Adage	64
33. Adage	66
34. Adage	68
35. Adage	70
36. Adage	72
37. Adage	74
38. Adage	76
39. Adage	78
40. Adage	80
41. Adage	82
42. Adage	84
43. Adage	86
44. Adage	88

Contents

45. Adage	90
46. Adage	92
47. Adage	94
48. Adage	96
49. Adage	98
50. Adage	100
51. Adage	102
52. Adage	104
53. Adage	106
54. Adage	108
55. Adage	110
56. Adage	112
57. Adage	114
58. Adage	116
59. Adage	118
60. Adage	120
61. Adage	122
62. Adage	124
63. Adage	126
64. Adage	128
65. Adage	130
66. Adage	132
Murks Behind Clear Water	135
Gratitude	145

Foreword

"Search for the transcendental level of knowledge."

-Keerthana Senthil kumar.

Preface

BOOK DESCRIPTION:

"PHILOSOPHICAL PHYSICS" is a book which compares Physics with Philosophical facts. It's,

Neither proverbs nor quotes,

Neither poems nor proses,

Neither myths nor truths,

Instead it is all which resembles day to day life with science. It not only covers physics but also other subjects too. Let's not detail it, you will get to know once you complete it.

Acknowledgements

MOTHER

TO MY MOM,

Here, the daughter of you is presenting a gift which has been written by her.

Keerthana, one of the daughters of you don't know or not eligible now to afford you an expensive thing. But she can present you a thing which can't be bought by money.

She became inspired by you and the result of that is this book.

She will be satisfied and fulfilled only if you congrats her though many did. This book will be honoured only if you read this mom.(N.JEYAKUMARI).

ACKNOWLEDGEMENTS

FATHER

Keerthana! Don't you worry?
    I smile...
    Keerthana! Don't you cry?
    I smile...
    Keerthana! Don't you fight?
    I smile...
    Keerthana! Don't you trust?
    I smile...
    How could you smile always inspite of these Keerthana?
    I smile... Don't you know why?
    Yes, how?
    You don't know cos you call my name wrong, call me KEERTHANA SENTHIL KUMAR...
    This is my name and
    the reason behind my smile is
    the person behind my name.

TEACHERS

Life is a communication system.

Mom and dad are the transmitting antenna who gave us birth and care,

Our goal is the receiving antenna where we finally reach,

Teachers are the electromagnetic waves who take us from this to that by encouraging and teaching.

(TO ALL MY TEACHERS).

ACKNOWLEDGEMENTS

GOD

Silence which can't be understood
    Tears which can't be wiped
    Memories which can't be erased
    Happiness which can't be yelled
    Truth which can't be explored
    Expectations which can't be done
    Etc............
    Etc............
    Are known only to the one
    That's GOD.
    Don't believe in GOD.
    Instead trust him.
    Everything will be fine....

Prologue

PHYSICS

It's not PRESSURE and TENSION.
It's TREASURE and INVENTION.
Learn it with PLEASURE and ATTENTION.

PHYSICS

P hysics is a natural phenomenON – if you
H eartily study, your creativity will ON
Y oung scientists will BORN
S oon unknown things will be GONE
I nterest and involvement gives IMPROVEMENT
C reative and knowledge gives DISCOVERMENT
S uch a thing is ACHIEVEMENT.

TO DREAM AS A PHYSICIST

Your eyes' convex lens should converge physics concepts and focus it to brain,
Your nose should smell future physicists for the welfare of the nation,
Your eyes should penetrate only positive energy of tranverse and longitudinal waves
Your mouth should send the frequency of Physics to make it audible,
Your hand should follow Flemming's right hand rule in a right manner,
Your leg should walk on the straight path in which LED light glows...
If you use these six organs correct, then as a Physicist,
You will enjoy the seven colors of joy, now as a prism...

PROLOGUE

MY LOVE...BOOK LOVE...

I look
But appearances are deceptive
Fail of love at first sight.
I took
Glimpse of hope
Pass my struggle.
I book
In sound of desire,
To gaze at silence of words.
I open
Into the paradise where,
One eye drops tears
One eye melts heart,
Some wit births
Many shits of deaths
I tempt
Inbetween the pages
I fall
In search of reality
And found loyalty.
I love you
Like waves striking the lines
More! And more! Again!
To own you one day
And that day will be right soon...

ONE
ADAGE

The anger in our face is the Sinusoidal wave,

But the smile in our face is the tangential wave.

PHILOSOPHICAL PHYSICS

ELUCIDATION:

When we are in anger, some used to stay calm or cry, but some used to shout at others or hurt them.So there is no use of anger which gains nothing but loss like the output of sinusoidal wave which is zero.

But anger is must in some places like where sinusoidal wave is very important in AC (Alternating Current).

When we speak of smile. Even when someone hurts you by scolding, shouting etc..Treat them with the peaceful smile which makes them regret about their behaviour.It's like the output of tangential wave which is infinity.

TWO
ADAGE

Life is the radius of curvature(R)

Principal focus is our ambition(F)

Then our focal length(f) life is happy.

ELUCIDATION:

Our life time is taken as radius of curvature, no matter whether it is for concave or convex lens / mirror.

Then principal focus is our ambition. The next half part of our life is get started after our point of ambition.

If we attain our ambition, then our focal length life will be too happy to lead.

Even in optics or our life, principal focus plays a vital role. So let's focus on our ambition.

THREE
ADAGE

―・♡・―

We are electromagnetic waves,

We can't propagate straight without

Parents of electric and magnetic field.

ELUCIDATION:

Here electric field is referred to as mother and magnetic field is referred to as father and direction of propagation is our life.

The electromagnetic wave does not propagate in the straight direction without the help of electric and magnetic fields.

Likewise, without obeying our parent's words we could not live a positive life.

FOUR
ADAGE

We should be the moment of inertia,

When we are at the situation of failure.

ELUCIDATION:

When a bus driver applies break, the bus stops but our body tends to move in the direction of motion.

This property is known as moment of inertia.

Likewise, when we get a failure, we should tend to move in the direction of success.

That's why, use the moment of inertia at the time of failure.

FIVE
ADAGE

---·♡·---

Creative thinking is not a convex lens.

It's actually a concave lens.

ELUCIDATION:

Convex lens is a lens which converges to a point which refers that our thinking should not be narrow.

Concave lens is a lens which diverges to many points which refers that our thinking should be broad.

Creativity develops only then we will achieve something different.

SIX
ADAGE

Success and failure are as like as

Couple in Physics.

ELUCIDATION:

Couple in Physics is the two equal parallel forces acting in opposite directions. These two forces don't have any direct contact.

Likewise, success and failure give two equal amount of feelings but it is opposite to each other.

The former gives happiness but the latter gives disappointment.

But we should treat them equally as a part of learning.

SEVEN
ADAGE

Don't waste time on youtube

Which oscillates you,

Instead use U-tube to find

Time period of oscillations.

ELUCIDATION:

If we start watching a video from you tube, we will sink in it and continue to the upcoming videos.

The youtube is very clever, it will always induce us to see it more by showing some interesting unwanted videos.

Instead by calculating time period of oscillations in U-tube, it makes you understand the value of time.

The only difference in these things is that, one makes the time waste for hours and the other makes it valuable for even a minute.

EIGHT
ADAGE

Always act like a buoyancy force.

ELUCIDATION:

In physics, buoyancy or upthrust force is an upward force exerted by a fluid that opposes the weight of an immersed object.

Here it is related to life, even though you have many pressures compressing you from outside, that is the problems or obstacles, you should always be the buoyancy force or upward force.

In which you should solve all the problems and pressure which compresses you down and come up.

NINE
ADAGE

In our life,

Everything happens as per

Newton's third law of motion.

ELUCIDATION:

Newton's third law states that for every action there is an equal and opposite reaction.

When you are just lazy and doing nothing, the action of lazy might give you happiness but the reaction will be suffering.

When you are working hard to attain your goal, the action may give you suffering but the reaction will be happiness.

TEN
ADAGE

Imagine your brain as a plug key

And work as a battery.

ELUCIDATION:

When you switch on the plug key, the battery starts working and we get some output due to its function.

Similarly, when you think of your goal in brain, that is, the switch is on, only then you can work for it.

Finally, you will get the output of success as glow.

ELEVEN
ADAGE

You should be an equipotential surface

At any kind of situation.

ELUCIDATION:

Any surface with same electric potential at every point is called an equipotential surface.

You should be like it, wherever you check, the potential will be same.

Likewise you should be strong in your attitude and never give it up at any circumstances.

TWELVE
ADAGE

Human body can be an example

For Conductor,

But human brain should be an example

For insulator.

ELUCIDATION:

Conductor, a body which allows charges to pass through. The perfect example is human body.

Insulator, a body which doesn't allow charges to pass through. The perfect example is plastic.

Here, the insulator is compared to human brain which never allow charges to pass through, that is, the unwanted thoughts to pass through.

THIRTEEN
ADAGE

Be a quantum mechanics which is

Different from other mechanics.

ELUCIDATION:

Quantum mechanics is the one which is beyond and different from all other mechanics.

We should be like quantum mechanics, that is,

We should think different

We should do different

We should observe deep and succeed in a different manner in which anyone cannot think of that extreme.

FOURTEEN
ADAGE

Love is like inserting the dielectric slab

Between the plates.

ELUCIDATION:

When a dielectric slab is inserted between the plates or capacitors, its capacitance increases but the potential decreases.

Likewise when a man fall in love, that is, love has entered into his life, his emotions increases but the knowledge decreases, then the life will be in vain.

FIFTEEN
ADAGE

All people in the world occur in integral multiples of this basic unit.

$q = ne$

ELUCIDATION:

It is the basic simple equation or formula in which

q is the charge,

n is the integral multiples,

e is the charge of the electron.

But here it is,

q is the people,

e is the angle in which they see other people,

n is either positive or negative angle.

SIXTEEN
ADAGE

---♡---

Cognition of teachers are the nuclear fission reaction.

ELUCIDATION:

Cognition – the mental action or process of acquiring knowledge and understanding through thought, experience and the senses.

Nuclear fission – it is the reaction of decay process in which the nucleus of an atom splits into smaller parts.

It is the chain reaction.

Likewise, the knowledge of teachers will split into the students and the students to others and so on.

It is also the chain reaction.

SEVENTEEN
ADAGE

The difference between heart and brain

Is that,

Heart is nerve fibre,

Brain is capacitor.

ELUCIDATION:

Nerve fibre – permeable to charges and also conducts.

Heart – permits everything inside and gets hurt soon.

Capacitor – impermeable to charges but it conducts.

Brain – looks before to leap.

EIGHTEEN
ADAGE

Growth of thymus in human

Is the path of projectile.

ELUCIDATION:

Thymus gland is one of the primary lymphoid organs.

It starts to grow during the foetal stage and continues growing and then reaches the maximum size at puberty.

After that its size reduces and becomes zero at death.

Which is like the path of the projectile and it is parabolic.

NINETEEN
ADAGE

If you retrospect and introspect

your activities,

You will clear the murky past.

ELUCIDATION:

When you clearly review and examine your activities you can correct your dirty past and lead a good future.

Everyone in the world will have some bad and dirty experience in their past.

To made it not happen in the future we should clearly notice each and every move done in the past life.

And should examine our mistakes and that not be committed again.

TWENTY
ADAGE

———♡———

Success is inversely proportional to

Osmosis.

ELUCIDATION:

Osmosis, it is the flow of liquid from the higher concentration to the lower concentration.

Reverse osmosis is the flow from lower concentration to higher concentration.

Here then, success is directly proportional to reverse osmosis and inversely proportional to osmosis.

To attain success, we should start from the lower place to the higher position.

TWENTY-ONE
ADAGE

One's attitude should be eurythermal

And not be stenothermal.

ELUCIDATION:

Eurythermal – *able to tolerate wide range of temperatures.*

Stenothermal – *able to tolerate only small range of temperatures.*

Our attitude should tolerate wide varieties of people in a bold manner like eurythermal.

But should not be broken even with few people's activities like stenothermal.

TWENTY-TWO
ADAGE

Don't worry that you're in the

Microstate level of knowledge.

Think that macrostate is the combinations of microstate.

ELUCIDATION:

***Microstate** – single instant.*

***Macrostate** – group of single instants.*

Knowledge should be grown from single lesson to plenty of lessons.

TWENTY-THREE
ADAGE

I lost myself

In the forest of pens,

The wild of books,

And the zoo of library.

ELUCIDATION:

Here I compared the trees in the forest as pens and wild animals as books and zoo as a library.

In zoo, wild animals from different forests has been collected and saved.

Similarly, in library, books from different places and authors has been collected and saved.

That is, I completely become paralysed when I go into the library and take a book and start writing using my precious pen.

TWENTY-FOUR
ADAGE

Don't be Amrita Devi,

Be the brave Jhansi Rani.

ELUCIDATION:

Amrita Devi is a pity girl who sacrificed her life for a tree in her place.

But the brave Jhansi Rani is a strong girl who opposes the opponent boldly in the war and defeat them.

TWENTY-FIVE
ADAGE

Our knowledge should be a sperm

And not be an ovum.

ELUCIDATION:

After puberty every man produces sperms and it continues till his death.

But to women after puberty, she utilizes every ovum and it ceases out at the middle.

Our knowledge should be like sperm which should be continued till death and not be an ovum.

TWENTY-SIX
ADAGE

An atom can never longlast in an excited state like our happiness.

ELUCIDATION:

An atom will excite to the higher energy state when it is subjected to external energy.

But it cannot longlast in the excited state, soon it will come back to ground state.

Similarly, the happiness in our life never longlasts till death.

Changes never change.

TWENTY-SEVEN
ADAGE

Brain should act as

Quantum mechanics,

Which is unusual among others.

And rememberance as

Classical mechanics,

Which is limiting case.

ELUCIDATION:

Quantum mechanics is a part of physics which is different from all others.

Quantum mechanics contains classical mechanics as a limiting case, yet at the same time it requires the limiting case for its formulation.

Here , Brain is compared to quantum mechanics. It contains rememberence as the limiting case. Because of remembering the sad past we are suffering in present. Still the brain requires this for its formulation. Because rememberence is the main tool for functioning our daily life.

TWENTY-EIGHT
ADAGE

A good person acts as a spherical shell.

ELUCIDATION:

Spherical shell is the one in which the charges inside it is zero but it has a potential.

Likewise, we should not allow the unwanted thoughts inside us and has a potential or knowledge or capacity to do work inside us.

TWENTY-NINE
ADAGE

No person is as loyal as book.

No friend is as loyal as parents.

ELUCIDATION:

We should not compare our friends with parents as no friend will be loyal as parents to you.

We should not compare our friends or parents or any person in this world with books.

As no individual in the world will be as loyal as book.

Books, which makes us gain a lot, expects nothing in return, never hurts or cheats or scolds at any cost.

THIRTY
ADAGE

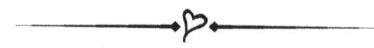

To reach certain destination

The one who choose distance, hardworks

The one who choose displacement smartworks

So let's be smart.

ELUCIDATION:

There is a difference between distance and displacement.

Distance is a scalar quantity which includes only magnitude.

But displacement is a vector quantity which includes both the magnitude and direction.

Displacement is the shortest distance but distance is the way we took, how long it can be.

One should be as smart as displacement by choosing their distance as displacement.

THIRTY-ONE
ADAGE

Life is a redox reaction,

One should act both as

Oxidizing agent and

Reducing agent.

ELUCIDATION:

Redox reaction is a reaction which is the combination of both the oxidation and reduction reactions.

The element which performs reduction reaction is called the oxidizing agent. The element which performs oxidation reaction is called the reducing agent. Here life is compared to redox reaction.

When you get failure, that is, reduction reaction, you should add your qualities and become oxidizing agent. When you get success, that is, oxidation reaction, you should reduce your headweight and become reducing agent.

THIRTY-TWO
ADAGE

Running water has more oxygen,

And running life has more obstacles.

ELUCIDATION:

Running water has more oxygen because the movement mixes the air into the water.

Likewise, our fastmoving life mixes with all our desires and leads to the increase in amount of obstacles.

THIRTY-THREE
ADAGE

You should never trust anyone,

But you should be trusted by everyone.

ELUCIDATION:

Trust , it's a big thing which plays a major role in our life.

Most of the people are betrayed only because of the trust they kept on others.

It's very better not to trust anyone in our life but you should understand the value of trust and lead your life as a trustworthy person for others.

THIRTY-FOUR
ADAGE

Being disloyal is like reduction reaction

And loyal is like redox reaction.

ELUCIDATION:

Reduction reaction is a reaction in which removal of oxygen occurs.

Redox reaction is a reaction in which both addition and removal of oxygen occurs.

If you are being disloyal to someone who trusts you, it's like removal of your oxygen which is called death. Without oxygen you are no more.

If you are being loyal to someone who trusts you, it's like adding and removing of oxygen, that is, inhaling and exhaling of oxygen makes you survive.

THIRTY-FIVE
ADAGE

Increase in mass and number of individuals are

Twin characteristics of growth,

Increase in interest and involvement are

Twin characteristics of achievement.

ELUCIDATION:

The twin characteristics of growth are its increase in mass and individuals.

When its mass increases and reproduce to give youngones is called growth.

The twin characteristics of aachievement are increase in involvement and interest.

When you are very much interested, you will automatically involve in it, then you will enjoy the fruit of achievement.

THIRTY-SIX
ADAGE

To attain success in life, you should follow the process of development of fruit in a plant.

ELUCIDATION:

Here to develop a fruit, ovules develop into seeds and their ovaries develop into fruit.

Likewise, to attain success in life, mistakes develop into experience and hardwork develop into success.

THIRTY-SEVEN
ADAGE

Be a gymnosperm which has naked seed,

But don't be an angiosperm which has seed inside the fruit.

ELUCIDATION:

Gymnosperms are the plants which have naked seeds like conifers and cycads.

Angiosperms are the plants which have seeds inside the fruits like apple and mango.

We should be a gymnosperm, that is, should be very frank and open-hearted to everyone.

We should not be an angiosperm, that is, should not be hollow-hearted and selfish.

THIRTY-EIGHT
ADAGE

Be haplontic,

Or be diplontic,

But don't be haplodiplontic.

ELUCIDATION:

Haplontic is the life cycle of an organism with haploid chromosomes.

Diplontic is the life cycle of an organism with diploid, that is, paired chromosomes.

Haplodiplontic is the life cycle of an organism which follows sometimes haplontic and sometimes diplontic.

We should be haplontic, that is, take a single decision very strong in life.

Or else, be diplontic, that is, have another opinion if the above one doesn't occur.

But we should not be haplodiplontic, that is, oscillating in both decisions without a clear view.

THIRTY-NINE
ADAGE

It's not self control

That

You didn't go bar and consume

It's self control

That

You be in bar without consuming alcohol.

ELUCIDATION:

One should not blame others for your mistakes or wrong path.

Everyone should have limits and self control on theirselves.

If we are strong enough with that, none can influence us wrong.

FORTY
ADAGE

Stem is the basic requirement

Not only for plants

But also for humans.

ELUCIDATION:

Stem, which is the part of the plant in the shoot system and without it, the plant cannot survive.

And here the STEM – Science Technology Engineering Mathematics are the very basic things without which humans cannot survive.

FORTY-ONE
ADAGE

Hard work is hard

When you do that without interest.

Hard work is pleasure

When you do that with passion.

ELUCIDATION:

The things which you are doing without any sought of interest or motive, it will be hell to you.

If we do anything with keen interest and passion, it will be like we are in the heaven enjoying such pleasure.

FORTY-TWO
ADAGE

———•♡•———

Our brain should act like a macrophage.

ELUCIDATION:

A macrophage is a type of phagocyte, which is a cell responsible for detecting, engulfing and destroying pathogens and apoptotic cells.

It is the one which destroys foreign organisms and give us self- defence.

Likewise, our brain should destroy our unwanted thoughts and should give us self-control.

FORTY-THREE
ADAGE

If you forget a single (-) sign in ten mark sum, the whole ten marks will be gone.

If you forget your parents in life, the whole life will be in vain.

ELUCIDATION:

In maths problems, each and every positive and negative signs are very important.

If we forget it even in a single place, the whole problem will go wrong and net result will also be wrong.

Likewise, in our life, if we forget our parents or disobey them, I am damn sure that the net result of our life will become a big tragedy.

FORTY-FOUR
ADAGE

Ambition is like fertilization,

One should act as sperm

And not as ovum.

ELUCIDATION:

In this fusion of gametes, the male gamete called sperm and the female gamete called ovum is involved.

During fertilization, there will be the entry of many male gametes inside the female body. The only one gamete which moves fast and touch the ovum with certain characteristics can form a foetus.

But the female gamete is just stable and allows the male gamete in for fertilization.

We should act as sperm which moves towards ovum to fertilize it. And should not be ovum which will not move towards sperm to get fertilize.

FORTY-FIVE
ADAGE

Ambition is an Archimede's principle,

To attain it, you must follow that principle.

ELUCIDATION:

Archimedes principle states that the upward buoyant force that is exerted on a body immersed in a fluid, whether fully or partially submerged is equal to the weight of the fluid that the body displaces and acts in the upward direction at the centre of mass of the displaced fluid.

Here, ambition is compared with Archimedes principle.

To get the actual mass of an object, you will insert the object whose weight is greater than its mass.

Similarly, to succeed in your ambition, you should work very hard beyond the level of success.

FORTY-SIX
ADAGE

Our brain should not be a spleen

But should be a lymphoid organ.

ELUCIDATION:

Spleen, it is an abdominal organ involved in the production and removal of blood cells in most vertebrates and forming part of the immune system. Simply, it is called as graveyard of red blood cells.

The main function of lymphoid organ is defence. Spleen is also one of the lymphoid organs.

But here it is compared to obstacles. Our brain should not be the graveyard of obstacles and negatives. It should be a lymphoid organ and defence every obstacles and turn into positives.

So don't dump any wastes into your brain instead defeat it.

FORTY-SEVEN
ADAGE

Brain is an ovary which should be utilized until it ceases out.

ELUCIDATION:

Ovary, a female reproductive organ in which ova or eggs are produced in humans and other vertebrates as a pair.

At the time of puberty, there will be nearly 6000 to 8000 ovary and it gets utilized each and every month at the time of menstrual cycle and finally it get ceases out at menopause stage.

Here, our brain is compared to ovary. It really means that everyone has more capacity and talents within them. It should be utilized and exposed at the correct time and should achieve more and more in their lives.

FORTY-EIGHT
ADAGE

Be a pollen grain of hardy exine

And thin intine,

As the pollen's exine and intine have

A strong mind and kind heart.

ELUCIDATION:

Pollen grain is made up of exine and intine.

The exine is very hard and intine is very soft.

Likewise, we should have very strong and brave mind with very kind and soft heart.

This makes our life healthy.

FORTY-NINE
ADAGE

If we calculate the bond order

Between us, parents and friends keeping us as antibonding electrons,

We can know the difference.

ELUCIDATION:

Bond order is the ratio of difference of number of bonding electrons and number of antibonding electrons to two.

If we calculate bond order between our parents and us, it will be positive.

If we calculate between our friends and us, it will be negative.

Since our friends are not more positive than us, infact they will have demerits than us except few.

So we should understand the bond of our parents, they are more powerful than our friends.

FIFTY
ADAGE

Don't use money as endothermic

Use it as exothermic.

ELUCIDATION:

Endothermic reaction is a reaction in which the heat is absorbed from outside surroundings to inside it.

Exothermic reaction is a reaction in which the heat is released to outside surroundings from inside it.

If we use money as endothermic, that is grabing from everyone around you, you will earn more money but not the true persons around you.

If we use money as exothermic, that is giving to anyone generously, you will earn a very good name and status among the true persons around you.

FIFTY-ONE
ADAGE

Relatives are cohesive force

But friendship is an adhesive force.

ELUCIDATION:

Cohesive forces are the forces of attraction between same materials.

Eg: water – water.

That is, relatives have some coincidence with us like blood relations, caste etc...

They will help us only because of their naming relations with us.

Adhesive forces are the force of attraction between different materials.

Eg: water – glass

That is our friends doesn't have any bond but still they will there for us when need.

FIFTY-TWO
ADAGE

All success will not give you satisfaction

But a single satisfaction is a real success.

ELUCIDATION:

Everyone has some sought of passion in certain goal and someone is very vigorous in attaining their ambition.

Though they failed in their ambition, they have certain will power and attain success in other field as there are many ways to succeed in life.

Eventhough they got success and earned more in other field, they will not get satisfaction.

But even if they didn't go to the higher position in their ambition or earned more money in their passionable ambitious field, they feel a very good satisfaction.

FIFTY-THREE
ADAGE

Your good qualities should be like

Meristimatic tissue and

Discipline as permanent tissue.

ELUCIDATION:

Meristimatic tissue is the one which divides continuously till the death of the plant.

Similarly one's good qualities should grow, develop and gets divided or increased till our death.

Permanent tissue is the one which stops dividing and remains permanent till the death.

Similarly, our discipline should attain certain stage and should be maintained till our death.

FIFTY-FOUR
ADAGE

Our friendship should be a

Mitotic division.

Our ego should be a

Meiotic division.

ELUCIDATION:

In a mitotic division, the genes are equally transferred from one generation to other generation.

Likewise, our friendship with our friend should be true and stable transferred till our death.

In a meiotic division, the genes are reduced from one generation to other generation.

Likewise, our ego with others should be reduced day by day.

FIFTY-FIVE
ADAGE

Be the gravitational force,

To be directly proportional

to the product of our parents

and inversely proportional

to the square of the negative persons.

ELUCIDATION:

We should be a gravitational force which is directly proportional to the product of the masses and inversely proportional to the square of the distance.

We should be directly proportional to what our parents say and should be loyal and respectful to them.

We should be inversely proportional to what those negative people backbite about us and ignore them to move on positively.

FIFTY-SIX
ADAGE

Gaining of knowledge is like

Filling of electrons by

Following aufbau principle.

ELUCIDATION:

While filling of electrons by following aufbau principle, after completely filled the 3s orbital, you must again go to the 2nd shell to fill 2d orbital.

Likewise while reading one chapter, you must recall the previous one which you have already read.

This method of recalling will make you gain knowledge that is you will remember for a long time.

FIFTY-SEVEN
ADAGE

Men are electrons

But women are protons

So being women

Be always trust worthy.

ELUCIDATION:

A man is considered as an electron here. If an electron (man) comes out of the atom (home), the atom will become positively charged and the electron go and join the other atom and gets satisfied.

A woman is considered as a proton. If a proton (woman) comes out of the atom (home), the atom will be collapsed.

So, understand the value of women.

FIFTY-EIGHT
ADAGE

Brothers are ionic bond

Friends are covalent bond

But sisters are co-ordinate covalent bond.

ELUCIDATION:

Ionic bond is the one which denotes the electron to others when needed.

Brothers are like ionic bond who will give anything for you.

Covalent bond is the one which shares it's electron to others.

Friends are like covalent bond who share everything with you.

Co-ordinate covalent bond is the one which attracts the shared pair of electrons towards itself.

Sisters are like co-ordinate covalent bond who will buy your things also.

FIFTY-NINE
ADAGE

Life is an atom model

Still now,

We can't define it's correct structure.

ELUCIDATION:

We know that many proposed many different models of atom.

Starting from J.J.Thomson, Rutherford and Neils Bohr model of atoms etc...

Still now we can't predict and conclude the structure of atom accurately.

Life is also like that, the next moment we don't know what will happen. We can't predict accurately.

Life is real and earnest.

SIXTY
ADAGE

No two electrons have identical quantum numbers in an atom.

No two persons have identical characters of all in the world.

ELUCIDATION:

Every electron in an atom has different quantum number and each has different works to perform. They might be identical in all quatum numbers but atleast differ in spin quatum number.

Likewise in our world no two persons are with actual same character.

Each one has some unique kind of talents and responsibilities to perform.

SIXTY-ONE
ADAGE

Friends are just alkanes

But parents are alkynes.

ELUCIDATION:

Alkanes are less reactive with only single bond and can be broken easily. It's formula is C_nH_{2n+2}.

Alkynes are more reactive with triple bonds and cannot be broken easily. It's formula is C_nH_{2n-2}.

Our friends cannot help in our life in certain extreme. They are limited. Just like alkanes can be broken easily.

But parents can do and go to any extreme just for us. Like alkynes, their bond cannot be broken easily.

SIXTY-TWO
ADAGE

To make your brain as neutron,

Subtract your atomic number

From mass number.

ELUCIDATION:

Here, the atomic no. refers to the negative thoughts and mass no. refers to the positive thoughts and neutron is the steady state of the brain.

Usually, we calculate the number of neutrons by subtracting atomic number to the mass number. To make brain steady all time, you must grow up your positive thought to that of the negative thoughts. Then only your brain will balance it.

You might have raised by the question that mass number is greater than atomic number then the remaining will be the positive thoughts. I mean this to make you understand the positive thoughts.

SIXTY-THREE
ADAGE

Friendship is a nucleon,

No one can break it except the very

Strong nuclear force called parents.

ELUCIDATION:

The nucleons are the particles which consist of proton-proton bonding, neutron-proton bonding, neutron-neutron bonding which cannot be broken by any of the forces like gravitational, electric, magnetic forces etc...

But there is only one force called nuclear force which is used in nuclear fission process to separate the nucleons.

Your hearty relationship with your friend is really true and unbreakable. But the relationship with your parents can defeat that friendship. This shows the power and love for parents.

SIXTY-FOUR
ADAGE

Life is a heterogenous mixture,

One should make it homogenous.

ELUCIDATION:

Heterogenous mixture is a mixture which is not purely mixed and it can also be separated.

Life is like that mixture which consists of happiness and obstacles and so on.

Homogenous mixture is a mixture which cannot be separated soon.

So, we ourselves must work hard to solve all the obstacles in life and should lead a confidential happy life without getting diverted.

That is, make the heterogenous into homogenous mixture.

SIXTY-FIVE
ADAGE

Be a negative charge

To attract positive energies,

And be a positive charge

To spread positive thinking.

ELUCIDATION:

Basically the symbol of negative charge is inward arrow and the symbol of positive charge is outward arrow.

We want to be a negative charge to attract all the positive energies around us and should increase our positive potential.

Then when we become more positive, we are like positive charge.

We should spread our positive thinking and make all around us full of delight positively.

SIXTY-SIX
ADAGE

Be as proton

Work as electron

Lead your life as neutron.

ELUCIDATION:

Be positive always like a proton which has positive charge and strongly bound with a great nuclear force in the nucleus that is, always be strong in your thoughts and action at any situation in your life.

Work like electron, since electron is always working inside the atom without rest and it is very important in the stability of the atom that is, our work should always be sincere at any situation and our part should be very important to a company to run it successfully.

Our life consists of many positives and negatives. We shouldn't take any of these two to the extreme and should be neutral always.

Murks Behind Clear Water

I opened the door of my lungs

Took a book from the shelves of my heart

Tears secreted right now

My first tears with the vision of you,

My first cry with the sound of you,

My first sick with the care of you,

I don't know how, why, where,

I fall for you...

But I know damn sure

I have been fallen only for you...

I yelled out, you're my love,

I am just a babe to fall for

I looked for cartoons,

But switched off to sight you.

I entered the play ground,

But stepped in to chase you.

I crossed my friends chatting,

But aparted myself to text you.

I was hungry,

But controlled to be fed with your love.

My relatives visited me,

But closed the door to have lovely time with you.

My dad called out for movies,

But made himself to hold love lamp.

A guy proposed,

But I knelt down to love you,

My pet pampered for love,

But I kissed you with love.

Everybody hated me,

Ignored me,

They called me dumb

They said me even to beg for something

She teased me behind,

He laughed at my attitude,

I smiled with all those stuff...

With unconditional love on you

I accepted all those stuff...

To live a long love life with you

My love was infinite

My love was weird

My love was in one side of physical balance

I was oscillating with efforts of love

I gave all my bleeding love to balance it

But you man! my man! my love!

Left me like torn slippers...

Left me like faeces in ditch...

Left me like chewing gums...

Sorry! Nope! You didn't.

Oh hell!!! I don't deserve you,

Oh hell!!! I loss my virginity,

I couldn't be accepted without you, with other

Will try to calm down

That's okay to erase you

Now, when I look back, it's like a blank paper.

In my 18 years of life,

Nothing is filled without you...

In my rest years of life,

How could I be filled without you???

Oh my baby! Please come back...

Oh my doll! I am confused...

There enters my crush,

He holds my hand tight and said,

"I am there for you."

He hugged me tight and said,

"We should not quit, cheer up!"

He looked at my eyes and gave a hope,

"I will be your man forever my girl."

Still can't be satisfied,

Tired of fake words

Yet can be moved on,

He gives me time,

He gives me success,

He gives me back all I lost at once...

He gives me a wreath of Queen.

He turned those stuff,

"Worth for nothing" into

"Deserves more than deserve"

Now,

Everyone loves me

Friends stay with me

My past changed as

Vice versa to positive...

I am an angel now!!!

Yupe! Crush can't fulfill

Nope! Love do...

This might be moved on

But memories can't be done...

There, my love end's without marriage.

The diary is closed with the name of,

LOVE : Mr. MEDICINE

CRUSH : Mr. PHYSICS

And now I am Mrs. Physics,

Who wished to be Mrs. Medicine...

Tears rubbed right now,

Shelves of heart get it back,

Doors of lungs closed again...

Breathe take - All love will not succeed but will give willpower to succed...

MURKS BEHIND CLEAR WATER

Gratitude

Thanking means a lot than anything.

I thank all the souls who crossed my life leaving an impact for this book.

A person who sowed the seed of physics in my heart....

-(Mr.V.PRABHAKARAN, my 7th std physics teacher, A.K.T Matric. Hr. Sec. School. Kallakkurichi.)

A person who poured water of physics concepts to that seed...

-(Mr.R.B.SABHIS KUMAR, my 9th std physics teacher. SOWDAMBIKAA Matric. Hr. Sec. School. Thuraiyur.)

A person who made the physics shoot rise from the seed without lying down instead standing high......

GRATITUDE

-(Mr.D.RAJARAM, my 12th std physics teacher. SOWDAMBIKAA Matric. Girls Hr. Sec. School. Thuraiyur.)

A person who made the shoot into varieties of branches of physics and other subjects too.....

-(Mr.T.HAREEIHARAN, my English professor. NEHRU Memorial College, Puthanampatti.)

The people who look after the tree without destroying with the care of physics water and manure....

-(Mr.R.KABILAN, Mr.P.RAMESH, Ms.P.BACKIALAKSHMI, my Physics lecturers. NEHRU Memorial College, Puthanampatti.)

The persons who found that there were also fruits in that tree.....

-(Mr.V.ABINASH, Ms.K.SATHYA, my best friends who encourage me for what I do.)

The one, only one who wished that fruit should be showcased in market to make other taste this Physics book.......

-(Mr.A. Michael, my brother who made me feel that I could do myself rather looking others.)

I can't end in a single word of "THANK YOU" to all these people of my life. I should be grateful for them at my life time. Each and every word and moment by them means a lot for this to be done.

Thanking all the readers for taking time for this book. It's nothing if you are not here. A tree will be valued by its lifetime. Here you are the age for my tree.

THANK YOU

Printed in the USA
CPSIA information can be obtained
at www.ICGtesting.com
CBHW031721241024
16327CB00023B/276